Revenge of the Sinister Universe:

The Reality of Everything

Greg Feild

September 5, 2018

About the author:

I earned a PhD in experimental high energy physics from the Pennsylvania State University working on HERA at DESY in Hamburg, Germany studying photoproduction and deep inelastic scattering in electron-proton collisions.

I did my postdoctoral studies with Yale University working at Fermilab on the CDF experiment at the Tevatron. My primary research interest was particle hadronization in charmonium production in proton-antiproton collisions.

Of miracles:

It will be sufficient to observe that our assurance in any argument of this kind [the testimony of men] is derived from no other principle than our observation of the veracity of human testimony, and of the usual conformity of facts to the reports of witnesses.

It being a general maxim, that no objects have any discoverable connexions together, and that all inferences, which we can draw from one to another, are founded merely on our experience of their constant and regular conjunction; it is evident, that we ought not to make an exception to this maxim in favour of human testimony, whose connexion with any event seems, by itself, as little necessary as any other.

Were not the memory tenacious to a certain degree; had not men commonly an inclination to truth and a principle of probity, were they not sensible to shame, when detected in a falsehood: Were not these, I say, discovered by *experience* to be qualities, inherent in human nature, we should never repose the least confidence in human testimony.

> -- David Hume
> *An Enquiry Concerning Human Understanding*

Abstract:

In this book, we continue to construct, critique, correct, and compactify our views on "The Sinister Universe"; last visited in the paper "On Rotation".

In particular, we propose two principles to explain all phenomena; the minimization of work, and the conservation of the three, classical and quantum mechanical, 'first integrals of the motion'; E, L, L_z.

perhaps there is only one principle -

action or motion

take your pick!

physics is fun!

Units and dimensions:

Heaviside Units, Heaviside Units (c=1), rationalized Heaviside-Lorentz units (c.g.s), rationalized Gaussian Units, Natural Units, . . .

Confounding and confusing!

In our last book, "On Rotation", we said;

Finally, in all our expressions and formulations we will bring all suppressed variables (i.e. c=1, h^{bar}=1) to the fore. To date, we've been carelessly mixing Units, Natural Units, etc.; obviously not without some unnecessary confusion!

Then, I lost my nerve, as I was still a bit uncertain, given the almost universal tendency of most authors to set c=1, sometimes without explicitly saying so; e.g. Best Book Ever! :(

Now, we state with confidence, the magnetic moment of the electron (Standard Model) is

$$\mu_e = (e\, h^{bar})/(2\, m_e\, c) \tag{a}$$

We will also, sometimes, find it useful to break the factor of 2π free from the factor h^{bar}. The origin of the factor 2π is for the conversion of frequency to angular frequency in the particle wave function. However, it also seems to have some significance as a space factor, representing the solid angle integral, or something like that. But, more on this later!

In our new, universal model, the magnetic moment of the electron is (6,13,14,18);

$$\mu = (e h^{bar}/2 m_e c)(1 + \tfrac{1}{2} v^2/c^2 + \tfrac{3}{8} v^4/c^4 + \ldots) \tag{b}$$

and the magnetic moment of the neutrino is

$$\mu = (h^{bar}/2c)(1 + \tfrac{1}{2} v^2/c^2 + \tfrac{3}{8} v^4/c^4 + \ldots) \tag{c}$$

Similarly, the electromagnetic coupling 'constant', alpha, is (14);

$$\alpha = \alpha_0 (1 + (v/c)^2 + (v/c)^4 + \ldots) \tag{d}$$

$$\alpha_0 = e^2/4\pi\varepsilon\, h^{bar} c \tag{e}$$

The gravitational coupling constant is

$$\alpha_G = (m_e^2 G)/(\hbar c)\,(1 + (v/c)^2 + (v/c)^4 + \ldots) \tag{f}$$

The weak coupling constant is

$$\alpha_W = (m_v^2 G)/(\hbar c)\,(1 + (v/c)^2 + (v/c)^4 + \ldots) \tag{g}$$

And finally, the strong coupling constant (14) is

$$\alpha_S = (G/4\pi\varepsilon)^{1/2}\,(2m_e e/\hbar c)\,(1 + \tfrac{1}{2} v^2/c^2 + \tfrac{3}{8} v^4/c^4 + \ldots) \tag{h}$$

In the universal model, all the 'constants' run, because *the fundamental coupling charge of a particle is the relativistic mass-energy* of the particle.

As particle velocities near the speed of light, the expansion terms in each of the four 'standard model' couplings will begin to dominate, eventually dwarfing the constant, or 'rest mass', terms.

The scale at which the four standard model forces 'merge', or become equivalent, will boil down to a matter of taste, dependent on one's choice of v/c.

In our model, there is no particle 'self-interaction', and no charge screening involved in the interaction between particles.

In our model, there are no fields, no vacuum, and no spacetime.

Never, ever, ever, has a virtual particle-antiparticle pair, ever appeared, anywhere, ever.

Not even 'near' a 'black hole'.

The mind of g-d:

a conservative accountant

The electromagnetic charge:

Interactions conserve spin, mass, and charge. In the universal model, we have generalized the concepts of electric charge and mass, to include all conserved quantities.

We define the electromagnetic charge to be

$$Q_{EM} = e \hbar / 2c \; \mathbf{s} \qquad (i)$$

and the mass charge to be

$$Q_{MASS} = m \hbar / 2c \qquad (j)$$

These quantities must be conserved at every 'vertex'.

The nature of these new charges are summarized in Table 1.

Force	Coupling constant	Conserved current	Rotation basis	Conserved charge
Electricity	e/m_e	Mass isospin	e, mu, tau	$e\hbar/2c \; \mathbf{s}$
Gravity	m_e/e	Charge isospin	e, υ_e	$m\hbar/2c$

TABLE 1: Table of coupling constants, conserved currents, and charges: Note: $m_e/e = m_\upsilon$

Introduction:

Our construction of the universal model is finally coming to a close in this seventeenth (!) book.

All physical phenomena can be accounted for, explained by, or represented as, the minimization of work, and the conservation of energy and angular momentum.

This formulation accounts for all classical and quantum interactions.

This model also explains the nature and behavior of the elementary particles, and the energy hierarchy of the three particle families.

let's do physics !

:)

Newton's laws:

In this section we present a synopsis of the 'universal' formulation, or expression, of Newton's laws, as proposed in our last book, "On Rotation".

Newton's first law:

If a particle does not experience a change in *angular momentum* relative to *any* arbitrarily chosen point in the universal reference frame, then the particle is considered to be free (i.e. there are no net forces acting on it).

For a two body system, if there is no change in the angular momentum of *either body* comprising the two body system relative to *any* arbitrary point (this 'excludes' the choice of points lying along the unit vector, **r**), then the particles are *not interacting*.

Newton's second law:

A particle that undergoes a change in angular momentum relative to our arbitrarily chosen point (i.e. the origin of our coordinate system) is said to experience a net torque, τ ;

$$\tau = dL/dt = \mathbf{r \times F} \tag{1}$$

where the force, **F**, is defined by

$$\mathbf{F} = d\mathbf{p}/dt = d(m\mathbf{v})/dt = m\, d\mathbf{v}/dt + \mathbf{v}\, dm/dt \tag{2}$$

For two body 'central force' motion, the torques experienced by each individual body relative to our chosen reference point, are equal and opposite;

$$\mathbf{r_1 \times F_1} = -\mathbf{r_2 \times F_2} \quad ; \quad |\mathbf{F}| = |\mathbf{F_1} - \mathbf{F_2}| \tag{3}$$

⇔

Newton's third law:

During a two body interaction, the two bodies will undergo equal and opposite changes in their respective '*actions*'; i.e. they will have equal, and 'opposite', changes in kinetic energy.

$$\delta \int dL/dt \cdot \omega \, dt = 0 \qquad (4)$$

where

$$\tau_{TOTAL} = \tau_{FORCE} + \tau_{SPIN} \qquad (5)$$

and

$$\tau_{SPIN} = l_1 \times B_2 + l_2 \times B_1 \qquad (6)$$

Newton's universal law of gravitation, expressed in the center of mass of a two body system, becomes;

$$F/E_{TOT} = K^*(c/R)^2 \mu - K^*(\mu v^2/R^2) - K^*(l^2/\mu R^3) \qquad (7)$$

$$K = G/c^2 \qquad (8)$$

where the second term on the right hand side of equation (7) is the *coriolis* force; our answer to spacetime disturbances.

Since our new coriolis force term goes as $1/R^2$, the classical and quantum mechanical conservation of the the first three integrals of the motion, E, L, L_z, is still guaranteed.

Conservation of energy and momentum:

The two algebraic relations of relativistic physics

$$E = T + m_0 c^2 \qquad (9)$$

$$E^2 = (pc)^2 + (m_0 c^2)^2 \qquad (10)$$

can only be realized using "complex algebra". In the Lorentz transformation derivation, the imaginary part arises or is due to a 'rotation of z about the axis ict'. In the Dirac equation derivation, we must employ the Pauli matrices which introduces the factor i.

The simultaneous satisfaction of equations (9) and (10) can also be achieved if we define the total energy to be a complex number

$$E = m_0 c^2 + ipc \qquad (11)$$

This should not be too shocking in this day and age as almost everything is imaginary or complex (e.g. wave functions).

The factor pc is the kinetic energy, as we shall see, and reduces to $p^2/2m$ in the nonrelativistic limit. The factor of i arises in a 'natural' way when we recast equation (11) as an operator

$$E_{OP} = (m_0 c^2 - i c \hbar \cdot \nabla) \qquad (12)$$

$$E^2_{OP} = (m_0 c^2)^2 + c^2 \hbar^2 \cdot \nabla^2 \qquad (13)$$

We can now define the kinetic energy operator

$$T_{OP} = -i c \hbar \cdot \nabla \cdot \qquad (14)$$

Elementary particles:

In our last book, "On Rotation", we demonstrated the energy and momentum of an electron can be written

$$L = \sqrt{3}\,\hbar/2 + I\omega \tag{15}$$

$$E = \hbar\omega_0 + I\omega^2 \tag{16}$$

$$\omega_0 = 2\pi c/\lambda_0 \tag{17}$$

where λ_0 is the Compton wavelength of the electron, which we also take to be the 'rest radius'. There is no longer a factor of ½ in the equation for the energy (16) as these are the relativistic expressions. We also need the following relations

$$\hbar\omega_0 = m_0 c^2 = I_0 \omega_0^2 \tag{18}$$

$$I_0 \omega_0 = \sqrt{3}/2\,\hbar \tag{19}$$

We define the moment of inertia $I(\lambda)$ to be

$$I = m\lambda^2/(2\pi)^2 \tag{20}$$

The same relationships hold for the muon and the tau.

Now, here comes the exciting bit. For particles at rest

$$I_e \omega_e = I_\mu \omega_\mu = I_\tau \omega_\tau = \sqrt{3}/2\,\hbar \tag{21}$$

We see that I has units of $\hbar$ or $L \cdot t$.

An elementary particle of constant velocity conserves the three first integrals of the motion; E, L, L_z. The component of angular momentum L_z is projected along (i.e. parallel with) the velocity vector or the direction of travel. The total angular momentum, L, precesses about the direction of travel at a constant frequency.

Particle families:

In the universal model, the processes of elementary particle production and decay are governed by the principle of the minimization of the work, or the minimization of the difference between the rest mass energies and the kinetic energies of the particles involved.

$$\delta(m - m_o) = 0$$

We hypothesize that when generating high energy leptons, at some point it becomes more expedient for 'Nature' to generate a muon rather than an electron with a velocity v/c ~= 1.

If true, where is this point? Is there a hard and fast rule governing when to choose the higher family particle, or is there some probabilistic indeterminacy involved?

In our model, the muon 'sheds' energy and spin in the form of its neutrino. The energy and spin shed ensures that the resulting virtual lepton propagator has spin = 0, and is *massless*.

This requirement places constraints on the energy and momentum of the initial and final states in muon (and tau) decays, and should help to explain the choice of the final state lepton from the three particle families in a particular decay, as well as some of the current mysteries surrounding lepton universality.

We can solve for this threshold, by assuming the muon rest mass is equivalent to the largest *allowable* relativistic mass of the electron.

$$m\_mu = m\_e/(1 - v^2/c^2)^{\frac{1}{2}} \tag{22}$$

$$(v^2/c^2)_{THRESHOLD} = 1 - (m\_e/m\_mu)^2 \tag{23}$$

A similar calculation will result in the velocity threshold between the mass of the muon and the mass of the tau.

We are not really clear on what "allowable" is: be it hard and fast or dependent on the particular dynamics of each interaction . . .

The leptonic table:

LEPTONS   ANTI-LEPTONS

electron	electron neutrino	PARITY ⇔	electron antineutrino	positron
⇐	CHARGE	MASS ⇅	CHARGE	⇒
muon	muon neutrino	PARITY ⇔	muon antineutrino	anti-muon
⇐	CHARGE	MASS ⇅	CHARGE	⇒
tau	tau neutrino	PARITY ⇔	tau antineutrino	anti-tau
⇔	mass isospin	charge isospin ⇅	mass isospin	⇔

TABLE 1: The leptons and their interrelations; or the kleptogenesis of the leptoquarks.

Any lepton can be 'generated' from any other by the appropriate applications of the parity operator, the mass isospin operator, and our newly proposed 'charge isospin' operator.

Using various combinations of the step up and step down operators of SU(2) and SU(3), plus the parity operator, we can write any quantum mechanical interaction current in terms of the 'fundamental' neutrino neutral current.

The Schrodinger equation:

Our theory is a gauge theory, We will call it the universal gauge theory.   :)

Our 'gauge invariant' solution, as last explored in "On Matter, Mass, and Motion", now looks like this;

$$\psi = \exp(im_0 c^2/\hbar\, t)\, \exp(-imc^2/\hbar\, t)\, \exp(i\mathbf{p}\cdot\mathbf{x}/\hbar) \qquad (24)$$

$$\psi = \exp(\,i(\mathbf{p}\cdot\mathbf{x} - (m-m_0)c^2 t)/\hbar\,) \qquad (25)$$

The relativistic Schrodinger equation is obtained by replacing the operator $p^2/2m_0$ with the operator p^2/m.

Kinetic energy:

$$pc = ch/\lambda \qquad (26)$$

$$ch/\lambda = ch\nu/v \qquad (27)$$

$$ch\nu/v = ch^{bar}\omega/v = Ec/v \qquad (28)$$

$$E = mc^2 \qquad (29)$$

A little algebra yields

$$E = pv = mv^2 \qquad (30)$$

math is fun!

Reverse pilot wave theory:

In our theory, particles follow a well defined path determined by the momentum vector. The average momentum of a particle is obtained by squaring the wave function in the usual way.

To obtain the instantaneous or "interaction" momentum at any given point, we must use the real part of the particle wave function.

$$\psi = \sin(px - Et)/\hbar \qquad (31)$$

$$\partial \psi / \partial x = p/\hbar \cos(px - Et)/\hbar \qquad (32)$$

We define the velocity operator

$$\mathbf{V} \equiv -i\hbar/m \, \partial/\partial x \qquad (33)$$

For a free particle, $x_0 = 0$, $v_0 = v$.

$$x = v_0 t \sin(px - Et)/\hbar \qquad (34)$$

$$p = mv_0 \cos(px - Et)/\hbar \qquad (35)$$

The reverse pilot wave idea is presented schematically in Figure 1.

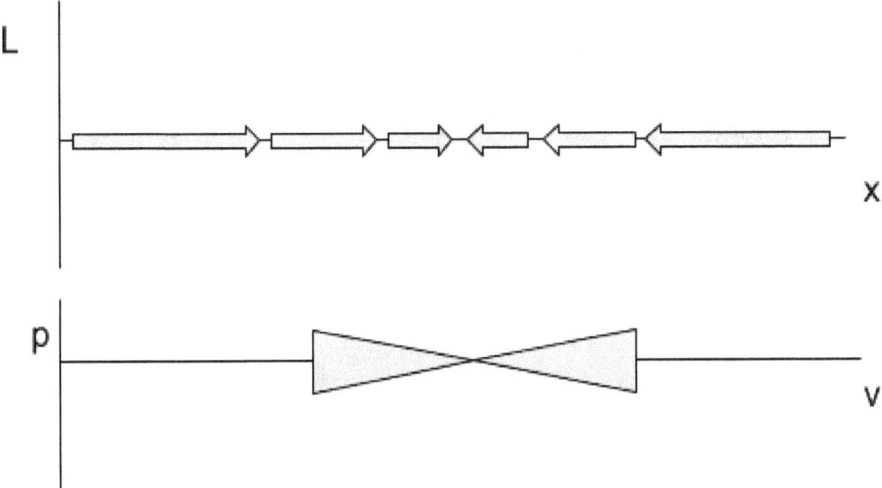

Figure 1: Reverse pilot wave theory.

The Dirac equation:

The universal Dirac equation is

$$H \psi = \alpha \cdot p \, \psi \tag{36}$$

$$i\hbar \, \partial \psi / \partial t = -i\hbar \, \alpha \cdot \nabla \psi \tag{37}$$

and is satisfied by, or solved with, the universal wave function, equation (25).

The hydrogen atom:

To date, we've been insisting that the electron orbits of the hydrogen atom are, or will be found to be, planar and circular. The straw man model. The actual orbits will be elliptical as determined by current quantum mechanics, however, the orbits are definitely planar.

Hydrogen orbits *must be* planar in order to conserve the constant first three integrals of the motion.

In our opinion, people determined an electron cloud, because they assumed, *beforehand*, the solutions would be spherically symmetric. One would not make such an assumption concerning the planetary orbits, although, it could be an interesting exercise.

In conclusion, the electron cloud is due to an honest mistake in judgement.

In the universal model, the hydrogen atom Schrodinger equation is the quantum analog of equation (7) which is our classical equation describing celestial orbits.

We also suggest the use of generalized coordinates, such as (θ, ω), and the corresponding uncertainty relationship

$$\Delta L_z \, \Delta \theta \geq \hbar/2 \tag{38}$$

Hydrogen atoms do not radiate in stationary states, because acceleration is only a necessary condition for radiation. Accelerated particles will only radiate, if there is also a change in *angular momentum*.

As an electron makes an orbital transition, or *accelerates*, it releases one unit of angular momentum, the photon, with the appropriate energy, etc.

We can now understand why classical electromagnetic waves are able to leave a source at the speed of light, while the sources themselves oscillate at much lower frequencies and speeds.

As one can see, our model explains *everything*, but spinning buckets !   :)

We don't do buckets.

Finally, the correspondence principle is easily explicable. The energy difference between photons emitted by electrons occupying very large orbits, becomes smaller and smaller, until the spectrum looks continuous and resembles classical electromagnetic waves.

However, emission is always discontnuous, and electromagnetic waves aren't real.

The proton:

In our model, the proton is a bound state of two positrons and an electron.

The strong coupling constant is (6)

$$\alpha_S = (G/4\pi\varepsilon)^{1/2} (2m_e e/\hbar c) (1 + \tfrac{1}{2} v^2/c^2 + \tfrac{3}{8} v^4/c^4 + \ldots) \qquad (39)$$

We imagine the orbits and description of the particles of the proton to be the relativistic generalization of the current model for the molecule H_2.

The proton is essentially antimatter !

On math and physics:

All phenomena can be reduced to the interaction of mass currents.

Indeed, in our model, the fundamental particle of matter, the lepton, is
a closed mass current spinning either to the left or the right. Particles spinning
in the same direction repel, and those spinning in opposite directions, attract.

A spinning, three dimensional lepton, has three planar projections of
spin angular momentum, **s • n,** resulting in an angular momentum per
unit area, $\hbar/2$, along each of the three geometric axes of space.

Electric charge is essentially a second mass 'scale' that allows for the
long distance interaction of the magnetic moments of two particles.

The point electric charge, e, is then the far field approximation of the
electromagnetic charge, or dipole moment; $e \hbar/2 c$.
The same holds for the mass.

In our model, there is no renormalization, no infitities, no singularities,
and no divergences; although we are not really sure what will happen with
ultrasoft photon emission and the 'infrared problem'.

We replace Q^2, with v^2/c^2, or $\Delta V^2 = (v_1 - v_2)^2$.

    The magnetic moment; a human-made, mathematical, contrivance,
    turns out to be the basis of all reality.

    Imagine that.

Reformation:

Are we not all tired of scientific revolutions?

Let us consider Reformation, instead.

Since "Copenhagen", the field of physics has been overwhelmed by dull-witted, fantasy prone, magical thinkers.

Depending on one's taste, the results have been either comic or tragic.

Or, of no consequence at all; because, who really cares?   :)

Conclusion:

Dear friends,

This is probably the last book we will write about physics for a while; of course, I think we've said this several times before!

I had a lot of fun, and I hope you did too.

Please take care of, care for, and care with, the universal model.

                                                      Enjoy!

                                                    Greg F

snake oil:

no embarrassment!
no shame

apologize not
or deign to explain

a physicist's credo
the psychics' creed

as priests
they steel
to the nave
turn heel !

to hide behind
sanctimony

duck blinds
of ceremony

still scrivening
their Impenetrable screeds

until they bleed

all coffers dry

questors heroic *!*

    questors heroic *!*
    tilt no more !

    toward
    the towering

    but towers !

    of ivory
    and babble

    bas relief
    in bronze

    canons and saints
    a chalice unknown

    pontiffs proclaiming
    ecclesiastically

    veneration

    of some dead
    philosopher's bone

I am:

    I am
    biology

    body and brain
    adrenaline
    bile

    sinew and sense
    pain trial

    homo sapien
    evolution's child

    reflex and reflection
    sight denial

    desire and doubt
    water loam

    a writhing

    wriggling
    flesh

    withering

    on recalcitrant bones

                            I am biology

                                            exploring itself

Resources:

Quantum Field Theory
Claude Itzykson, Jean-Bernard Zuber

Atomic and Quantum Physics
H. Haken, H.C. Wolf

Modern Elementary Particle Physics
Gordon Kane

Classical Dynamics of Particles and Systems
Jerry B. Marion

Foundations of Electromagnetic Theory
John R. Reitz, Frederick J. Milford, Robert W. Christy

Quantum Physics
Rolf G. Winter

Gauge Theories in Particle Physics
I. J. R. Aitchison and A. J. G. Hey

Quarks and Leptons: An Introductory Course in Modern Particle Physics
Francis Halzen, Alan D. Martin

Quantum Field Theory
F. Mandl, G. Shaw

Theoretical Mechanics of Particles and Continua
Alexander L. Fetter, John Dirk Walecka

and

Elementary Modern Physics   (Best Book Ever!)
Richard T. Weidner, Robert L. Sells

Books by Greg Feild: The SInister Universe Series

the pentateuch

1. "A quantum mechanical theory of gravitational interactions"
   CreateSpace Independent Publishing, 8/29/2016

2. "Observations on the quantum mechanical nature of gravity"
   CreateSpace Independent Publishing, 10/8/2016

3. "On gravitation and electric charge"
   CreateSpace Independent Publishing, 10/29/2016

4. "On spin, mass, and charge"
   CreateSpace Independent Publishing, 11/29/2016

5. "On angular momentum, acceleration, and absolute motion"
   CreateSpace Independent Publishing, 1/1/2017

the exegeses

6. "The Sinister Universe"
   CreateSpace Independent Publishing, 3/1/2017

7. "On Parity and Isospin"
   CreateSpace Independent Publishing, 4/11/2017

8. "Reflections on the Sinister Universe"
   CreateSpace Independent Publishing, 5/12/2017

the hermeneutics

9. "On Current Physics"
   CreateSpace Independent Publishing, 6/11/2017

10. "A Critical Examination of Classical and Quantum Mechanical Waves"
    CreateSpace Independent Publishing, 6/18/2017

the gospels   :)

11. "On wave particle duality and the quantum of action"
    CreateSpace Independent Publishing, 7/6/2017

12. "On matter, mass, and motion"
    CreateSpace Independent Publishing, 9/14/2017

13. "On action and reaction"
    CreateSpace Independent Publishing, 9/24/2017

14. "A quantum mechanical theory of everything"
    CreateSpace Independent Publishing, 11/5/2017

the compilations

"The Universal Model of Our Sinister Universe: The First Ten Books"
CreateSpace Independent Publishing, 7/2/2017

"The Canons of the Sinister Universe:
The Last Four Books on the Universal Model of Our World"
CreateSpace Independent Publishing, 11/5/2017

the expositions

15. "On Interaction"
    CreateSpace Independent Publishing, 4/21/2018

16. "On Rotation"
    CreateSpace Independent Publishing 8/19/2018

Learning physics from your dog:

Dogs know all about gravity. Dropped food falls to the ground.

Dogs also understand acoustics. The sound of plopping food will bring a dog running from rooms away.

If only I could teach my dog trigonometry.

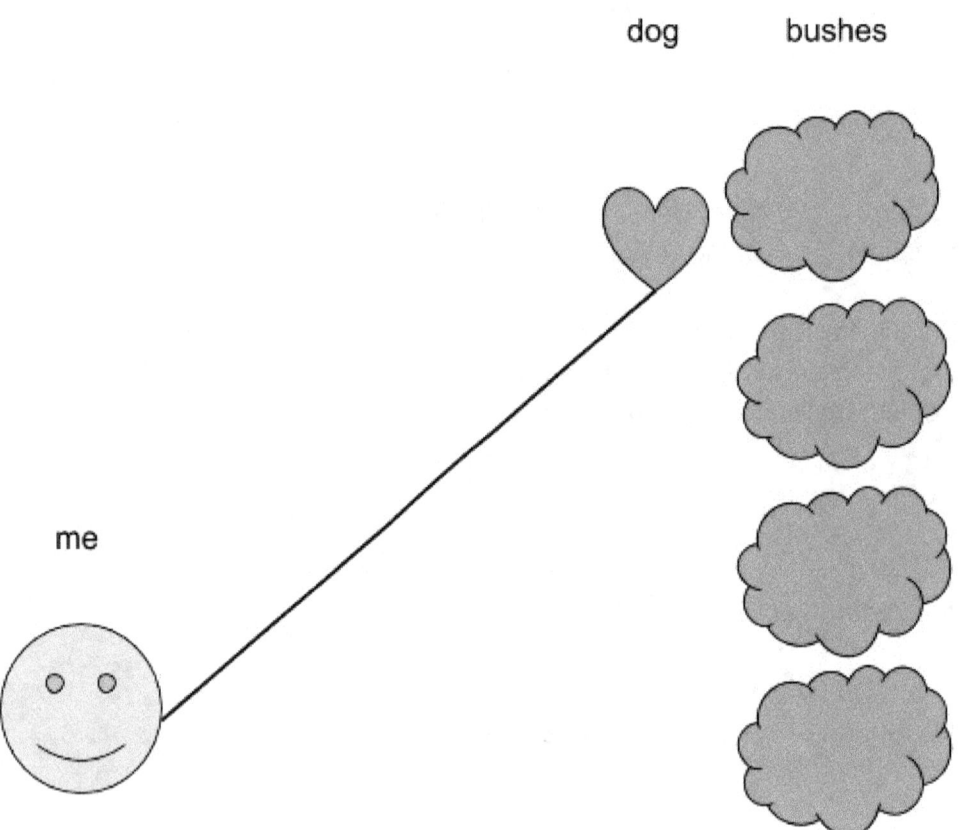

math is hard

www.ingramcontent.com/pod-product-compliance
Lightning Source LLC
Chambersburg PA
CBHW062345220526
45469CB00008B/2846